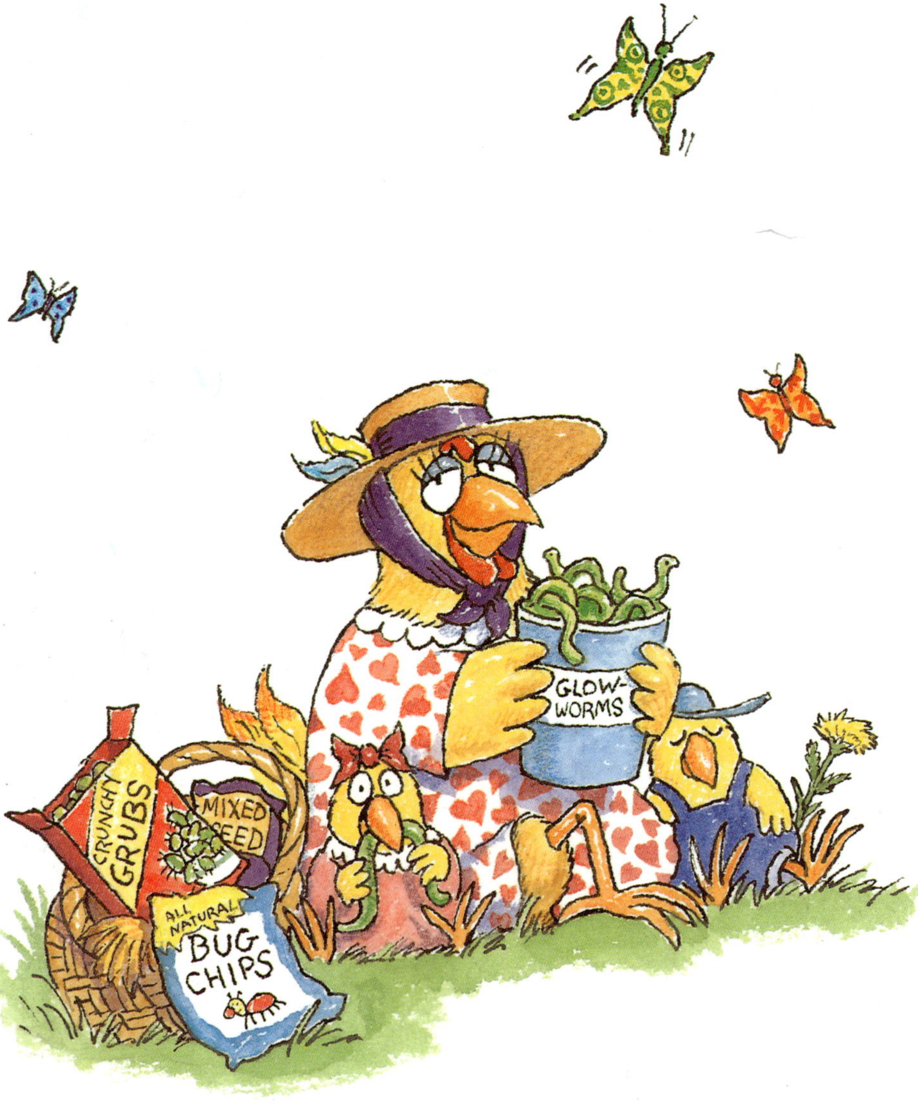
GLOW-
WORMS
CRUNCHY
GRUBS
MIXED
SEED
ALL
NATURAL
BUG
CHIPS

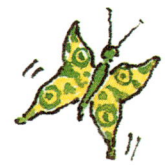

CHICKIE RIDDLES

by Katy Hall and Lisa Eisenberg

pictures by Thor Wickstrom

PUFFIN BOOKS

PUFFIN BOOKS
Published by the Penguin Group
Penguin Putnam Books for Young Readers, 345 Hudson Street, New York, New York 10014, U.S.A.
Penguin Books Ltd, 27 Wrights Lane, London W8 5TZ, England
Penguin Books Australia Ltd, Ringwood, Victoria, Australia
Penguin Books Canada Ltd, 10 Alcorn Avenue, Toronto, Ontario, Canada M4V 3B2
Penguin Books (N.Z.) Ltd, 182-190 Wairau Road, Auckland 10, New Zealand

Penguin Books Ltd, Registered Offices: Harmondsworth, Middlesex, England

First published in the United States of America by Dial Books for Young Readers,
a division of Penguin Books USA Inc., 1997
Published in a Puffin Easy-to-Read edition by Puffin Books,
a member of Penguin Putnam Books for Young Readers, 1999

3 5 7 9 10 8 6 4

Text copyright © Katy Hall and Lisa Eisenberg, 1997
Illustrations copyright © Thor Wickstrom, 1997
All rights reserved

THE LIBRARY OF CONGRESS HAS CATALOGED THE DIAL EDITION AS FOLLOWS:
Hall, Katy.
Chickie riddles / by Katy Hall and Lisa Eisenberg ; pictures by Thor Wickstrom.
p. cm.
ISBN 0-8037-1778-4.—ISBN 0-8037-1779-2 (library)
1. Riddles, Juvenile. 2. Chickens—Juvenile humor. [1. Chickens—Wit and humor.
2. Riddles. 3. Jokes.] I. Eisenberg, Lisa. II. Wickstrom, Thor, ill. III. Title.
PN6371.5.H3476 1997 818'.5402—dc20 94-33170 CIP AC

Puffin Easy-to-Read ISBN 0-14-130430-8
Puffin ® and Easy-to-Read ® are registered trademarks of Penguin Books USA Inc.

Printed in Hong Kong

The full-color artwork was prepared using pen and ink, colored pencils, watercolor, and gouache.

Except in the United States of America, this book is sold subject to the condition
that it shall not, by way of trade or otherwise, be lent, re-sold, hired out, or otherwise
circulated without the publisher's prior consent in any form of binding or cover other
than that in which it is published and without a similar condition including
this condition being imposed on the subsequent purchaser.

Reading Level 2.0

*Dedicated to
Chicken Little and Henny Penny*
K.H. and L.E.

To Sylvie and Sosha
T.W.

What is Snow White's brother's name?

Egg White—get the yolk?

If you crossed a cocker spaniel, a French poodle, and a rooster, what would you get?

A cockerpoodledoo!

What looks just like half a chicken?

The other half.

Where do hens come from?

Hennessee.

Where do chicks come from?

Chickago.

Why did the farmer hire the chicks to fix up the coop?

They were *cheep* workers.

Which side of a chicken has the most feathers?

The outside.

Why did the chicken cross the road?

To get to the other side.

Why did the chicken go halfway across the road?

She wanted to lay it on the line.

What do you get if you cross a road with a chicken?

To the other side.

Why did the turkey cross the road?

To show he wasn't chicken.

Why do hens lay eggs?

Because if they dropped them, they'd break.

Where do chickens go to dance?

To the Fowl Ball.

How do hens and roosters dance?

Chick to chick.

Can a chicken really be worth $2,000?

Only if she saves all her money.

What kind of weather
do chickens like best?

Fowl weather.

What holiday
do roosters love?

Feather's Day!

Why do chickens think cooks are mean?

They beat eggs.

What game do baby chicks love to play?

Peck-a-boo!

When is dinner
in the henhouse?

At eggs-actly seven o'clock.

What's made of chicken and noodles and can leap tall buildings in a single bound?

Chicken Soup-erman!

What's Chicken Soup-erman's real name?

Cluck Kent.

What day do chickens hate most?

Fryday.

What do you have to know to teach a chicken tricks?

More than the chicken.

What do chickens say
when they want
to trade nests?

"Let's make an eggs-change!"

Why did the hen take a hot bath?

She wanted to lay a hard-boiled egg!

Why did the hen stop laying eggs?

She was tired of working for chicken feed!

What do you call a chicken that likes to clean?

A feather duster.

What do you get if you cross a dog and a chicken?

A pooched egg!

What do you get
if you cross a chicken
and a police officer?

A hen that really lays down the law!

Have you heard the joke about chicken pox?

Shhh! Don't tell it! It might spread!

What's the opposite of cock-a-doodle-do?

Cock-a-doodle-don't!

Why didn't the chicken look both ways before she crossed the road?

She was a dumb cluck!

Why did the dinosaur
cross the road?

There weren't any chickens
back then!

How many chickens can you put into an empty coop?

Just one. After that it isn't empty!

Why should you never tell a joke to an egg?

Because it might crack up!

Why did the chicken run away?

She felt cooped up!

Why did the hen run away from the mall?

She heard it was a chopping center!

Why did the farmer invite his two new chickens to a party?

He was trying to make both hens meet!

How do we know
a rooster loves his comb?

Because he'll never part with it!

What's the favorite game in the henhouse?

Chickers.

What kind of jokes
do chickens like best?

Corny ones!